The Body

10 Things You Should Know

DR DARRAGH ENNIS is a professional quizzer and scientist. He has researched ecology and animal behaviour at Maynooth University in Ireland, outbreak dynamics of forestry insect pests on maple syrup plantations in Canada as well as brain development and RNA biology at both the University of Oxford and the University of Glasgow. In recent years he has also worked as a quiz professional on the ITV game show *The Chase*.

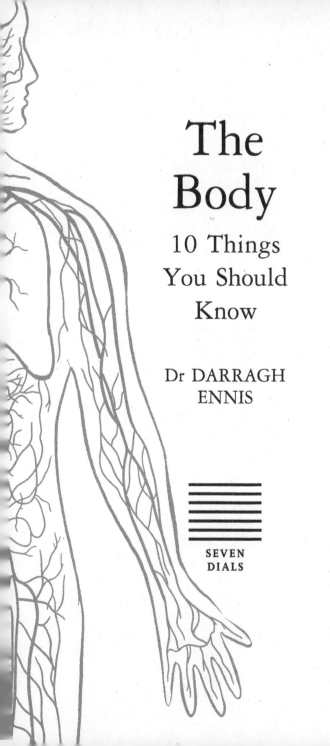

The Body

10 Things You Should Know

Dr DARRAGH ENNIS

SEVEN
DIALS

First published in Great Britain in 2024 by Seven Dials,
an imprint of The Orion Publishing Group Ltd
Carmelite House, 50 Victoria Embankment
London EC4Y 0DZ

An Hachette UK Company

1 3 5 7 9 10 8 6 4 2

A CIP catalogue record for this book Is
available from the British Library.

ISBN (Hardback) 978 1 3996 2627 9
ISBN (eBook) 978 1 3996 2628 6
ISBN (Audio) 978 1 3996 2629 3

Typeset by Born Group
Printed in Great Britain by Clays Ltd, Elcograf S.p.A.

www.orionbooks.co.uk

For Joan

Contents

Preface   1

1. From One Cell to Thirty Trillion   3

2. Why Does Our Body Age?   15

3. A Guided Tour of the Body on a
   Red Blood Cell   27

4. How Does Our Body Defend Itself?   37

5. Why Do We Sleep?   49

6. How Has Evolution Shaped Our Body?   61

7. How Does Our Body Get
   and Use Energy?   71

8. Feeling Hormonal?   83

9. It Just Makes Sense –
   How Our Body's Senses Work   95

10. The End? What Happens to
    Our Body When We Die?   109

Acknowledgements   119

Preface

Your body is miraculous. Yes, your body. The fact that you are sitting there in that chair, holding this book is miraculous. Or maybe you're reading in bed, or on a bus, or even in a bookshop, reading these lines and wondering if you should buy this book. No matter where you are you should understand that your body is a miracle of evolution and biology. The very fact that tens of millions of years of trial and error has led to a person who even understands what the word 'miraculous' means is nothing short of a miracle. We, each and every one of us, exist in a body that is so mind-bendingly sophisticated and yet we rarely, if ever, take the time to appreciate it. The aspects of our body that we notice and perceive barely skim the surface of what is going on inside ourselves, a complex series of interacting mechanisms that allow us to be who we are. Yet, ironically, this disconnect is our body making us focus on what it needs us to do, consume food for its energy, sleep when it needs to rest. So, while we may feel we are in the driving seat, we are also controlled and driven by our body's requirements. Even this 'we' who is being driven by our body is really only the culmination of the evolution of a sophisticated nervous system to form a brain that is self-aware.

Our bodies extend far beyond our conscious mind, no matter how much that part of our brain likes to think it's in charge. We all grow and develop, heal, move, consume food for energy and grow older. How all of these processes happen and are controlled is truly fascinating and, in this book, I hope to take you on a journey through yourself and give you a better understanding of what is going on in your body while you read it.

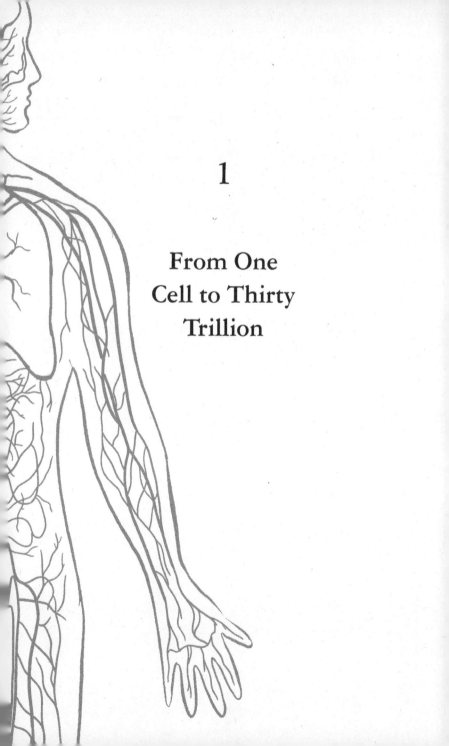

1

From One
Cell to Thirty
Trillion

What do you, the last person you spoke to, the winner of the Best Supporting Actress Oscar last year and Genghis Khan all have in common? You may not think anything at all (though after reading this book you may change your mind on that), but you definitely share one key thing. You all have one thing in common with every human on earth; you started out as a single cell. The formation of our body from that cell is a fascinating process, and even more so when you consider the starting material. One cell about the size of a pin head contains the spark of life and the ability to grow a complete human (with significant help from its mother, of course). That this spectacular transformation happens with about three gigabytes worth of information encoded in that cell's DNA makes it even more remarkable. But the key question is, how do we grow from such tiny beginnings to being the person holding this book on a crowded bus?

When the largest human cell, the ovum, fuses with the smallest human cell, a sperm, it forms what is called a zygote. This single cell is a genetic mixture of both parents, but it does not resemble any of the cell types we would recognise in an adult. There is almost no sign of the liver, skin or muscle cells we will find later. The concepts of moving, eating, thinking or even thinking about thinking are not part of what makes a zygote tick. What this single cell is concerned with is not being a single cell anymore. In order to grow a human, we need a lot more cells, about thirty trillion or so. The zygote needs to divide. And for the next two days or so, that's exactly what it does. In an adult, cells that are involved in growth etc. will divide about once a day and some others will do so much less frequently, or not at all. At the very beginning of our developmental journey however we need to increase cell numbers and fast. So, in these first forty-eight hours after the zygote is formed, one cell becomes thirty-two. This rapid division doesn't leave much time for anything else, including growth, so this group of cells is still the same size as the zygote and is known as a morula. While these cells are now much smaller than the zygote, they are still significant as they are very powerful. This is the final point of a body's development during which its cells are totipotent, meaning they can give rise to every cell type needed to make a body as well as the placenta, the tissue required to link the growing body to its mother's circulatory system so she can provide everything the body needs to develop and grow. Totipotent cells might not be familiar to you, but you may recognise them if I tell you they are a class of cell that can give rise to other types, stem cells.

From One Cell to Thirty Trillion

We will take a brief step away from our body's development journey to explain what stem cells are. For many people, stem cells have come to be noticed because of their potential in the treatment of diseases, perhaps as a way to grow new organs or tissues. Others may know of the ethical challenges of using these cells and acquiring them for research. But few know what they actually are. Put simply, they are cells that when they divide, don't simply make copies of themselves. They have the ability to divide into cells that have different functions. Some stem cells, like those in the morula, can form all of the cell types a body needs as well as external structures, such as the placenta, and are totipotent. More common during a body's early development are those that can divide to form any cell type in the body and are known as pluripotent. Pluripotent stem cells have a relatively short lifespan in our body and are absent in the adult, though scientists have managed to reprogramme non-stem adult cells in the lab to act like pluripotent cells for research; these are known as induced pluripotent stem cells. The final type of stem cell found in our body are known as adult stem cells, which are found (you guessed it) in our body throughout its life. Adult stem cells are mostly used to repair tissues and, usually, are only capable of producing a small number of cell types when they divide. Adult stem cells often undergo prolonged periods without dividing and are 'woken up' to divide when needed, e.g. to repair an injury.

Whatever their type, all stem cells have some things in common. The first is all stem cells can self-renew. Many of our cells, such as blood cells and nerve cells, do not usually divide and replicate. Stem cells are different in that they are capable of dividing many times. When a stem cell does replicate there are three possible outcomes:

1. Both cells are stem cells, increasing the number of stem cells in the body. This happens most commonly early in the body's development.

2. One cell remains a stem cell while another now has a different form and function (known as differentiation).

3. Both cells are more differentiated. This means the number of stem cells in the body decreases, something that happens more commonly in the body as the development journey continues.

The control of these three outcomes, in both location and over time, is of vital importance to the production, and maintenance, of a fully functioning and healthy body. Any deviation from the pattern of stem cell function can be catastrophic to the body, leading to improper development or even uncontrolled division of cells, causing tumours to be formed. The control of stem cells is very closely regulated in our body, with the turning on and off of certain genes at specific times helping to regulate this key process.

But back to the cluster of totipotent stem cells, still no larger than the head of a pin: the morula. From this point on, the developing body needs to increase in size as well as increase the number of cells. It also needs to begin forming structures that are more familiar to us, with a variety of cell types to carry out all the functions our body needs. To begin this process, the dividing cells begin to have different functions. The outermost layer of cells, which go on to form the placenta and other structures, now surround the cells that will form the embryo, which is sandwiched in between two fluid-filled cavities. One cavity is the primary yolk sac, and yes humans also have a yolk, which helps to feed the developing body while the placenta is growing into place. The other cavity will eventually go on to form the amniotic sac, a fluid-filled sac that helps protect the growing body in the womb.

From now on the overall body plan begins to take shape. This is an extremely complex process, but it can be reduced down to the body making tubes. Our bodies (and those of most animals that we think of as animals) have three distinct sections that are called 'germ layers'. These are essentially tubes – our digestive tract and a series of tubes wrapped around it. In order to make these tubes, the mass of cells forms a line called the primitive streak across them. The remarkable thing about that, is that this forms the midline of the body, and it is at this very early stage that the left and right side of our bodies is determined. The cells fold along this line (which can still be seen on parts of our adult bodies, such as the philtrum that runs from our nose to our upper lip) and form the tubes that make up our three body layers. The body now has distinct 'ends' at this point, the head and tail end are now set. A key player in knowing which end is which (known as patterning), is one of the first tubes to form: the notochord. The notochord is key to our body developing the right tissues in the correct places and it is also notable for its role in how we humans classify animals. All humans, fish, birds, reptiles and quite a few tiny sea animals all belong to one large group or phylum, the Chordata. The presence of a notochord at some stage of their development is a key qualification of belonging to this phylum, and it is at this early stage that our bodies possess one. Later in our development it goes on to form other tissues, most notably the discs in between our vertebrae. But for now, it helps to make sure our cells form up in the right places.

At this point, our body begins to become more recognisable, and the first organ to develop is a heart, which starts beating around five weeks into our development. It's the first organ to form and start functioning, closely followed by others, such as those in the digestive tract. The digestive tract and lungs form when one of the layers of cells rolls into a tube, with different sections forming small pouches that then develop into the stomach and intestines, with a small respiratory bud going on to form the lungs. Soon we will have a very familiar body, with growing and developing systems that are similar to our adult versions. In fact, almost all of the organs our body needs are formed within the first three months, though the brain and spinal cord continue to develop far past this point. With most of the body's systems in place, the body now begins to focus less on differentiation and more on growth. Our stem cells are now heavily outnumbered by the differentiated cells they have created, and most are no longer pluripotent. Instead, many stem cells are now dedicated to producing a small number of cell types that allows our body to grow. This is a process that continues long after we are born, and for some cells it happens throughout our lives, until we end up with approximately thirty trillion cells.

This change from one cell is astonishingly complex,

and a key question to ask is, how do the cells know what to do? During all of this growth and development, the cells are directed by the genetic information encoded in their DNA. Most people know very little about DNA, but essentially it is like an instruction manual on how to build a human. This DNA 'book' contains about twenty-five thousand genes which we can think of as pages. Each gene/page contains the instructions for a protein. Proteins are not just dietary supplements, they are essential to how our bodies function at a cellular level. If you want to build new cells, grow or carry out almost all functions in the body, proteins are needed. When a cell needs a protein, it uses the instructions contained in the relevant gene to assemble it. In order to do this efficiently, our cells don't move the massive DNA molecule around the cell (it's almost two metres long when stretched out!) so instead they make copies of only the relevant section, known as messenger RNA, which can be used as an instruction manual to make the proteins needed for growth and cell function. RNA is not just a messenger however, and the concentration of and interaction between these RNA and proteins helps to control how our cells and tissues develop. It is this control of how, where and when genes are expressed that is at the heart of how our cells 'know' how to develop into a fully functioning body. So even though almost all of our cells contain exactly the same DNA, how they use those instructions is key to our development. It is this 'interpretation' of our genetic instruction manual that allows us to grow our body from one cell to thirty trillion.

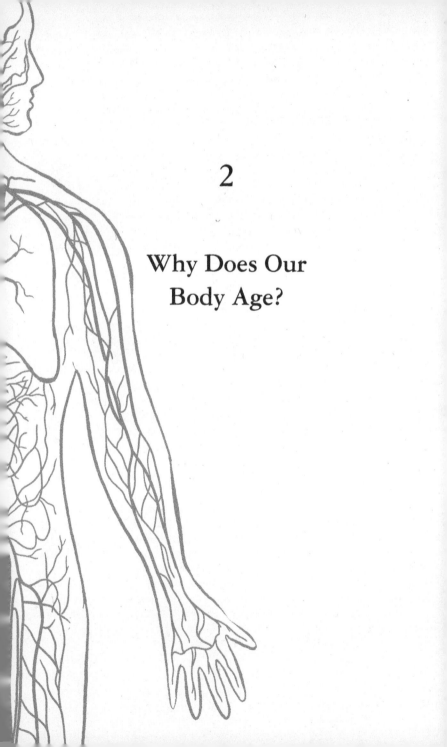

2

Why Does Our Body Age?

Have you ever been asked: 'If you could be an animal, which one would you choose?' Most people choose an animal with some amazing ability, like a high-flying and far-seeing eagle or a powerful lion. While these are not bad choices, they have missed possibly the most remarkable ability any animal possesses; they should choose the jellyfish. And not just any jellyfish, but specifically *Turritopsis dohrnii*. At first glance, this tiny jellyfish, just four millimetres across, may seem to be a ludicrous choice. Why would I pick this little blob instead of, say, a giraffe or a humpback whale? Well once you know the common name of *T. dohrnii* it becomes obvious; this little blob is known as the immortal jellyfish. For almost all lifeforms, aging of their bodies happens in one direction. 'Reversing the aging process' is strictly in the realm of fantasy and advertising slogans, our bodies get older every day. The immortal jellyfish however has managed to achieve what no amount of creams or supplements can do for us. *T. dorhnii* start life as fertilised eggs, similar to our own bodies. They then become a free-swimming larval stage, which find a hard surface to land on and form a polyp stage. This will grow and bud off as an adult medusa form, the familiar floating jellyfish form with tentacles. When an adult grows older, or is damaged or stressed, it is able to revert to the polyp stage and grow again into a new adult form. This is as if a butterfly can revert to being a caterpillar form again. This reversing of aging is, in theory at least, unlimited, making this remarkable body immortal.

Unfortunately for our human bodies, aging is something we all have to accept is a one-way process. Our bodies – from the moment they begin as one cell, grow and develop – also age. Our hair turns grey, our skin sags and wrinkles, our metabolism changes and all of our tissues and cells age. One real question we can ask is: 'Why do we age?' It is such a simple question, but if you learn one thing about our body from this book it's that the answers are often more complicated than the questions. One simple way our bodies age is down to simple wear and tear. Cartilage in our joints gets worn down by friction caused partly by constant movement, our skin gets cut and damaged by sunlight causing it to form scars, wrinkles and freckles. Less obvious and perhaps a more critical question to ask is: 'Why do our cells age?', as this is one of the root causes of our bodies growing older.

Like a lot of things in our body, the answer is down to our DNA and, more specifically, how our DNA copies itself when a cell divides. When a cell is ready to divide, it is critical that there are two copies of the DNA instruction manual so each cell can have their own. To do this, each of the chromosomes in the twenty-three pairs in the cell nucleus separate, and copies are made so there are now two sets of twenty-three paired chromosomes. Each set is then taken into one of the two new cells when division happens, so both new cells have a full copy of the DNA that the original cell had. Except it's not quite a full copy. Chromosomes are long, thin and threadlike, and, like a piece of thread, they have ends. In order to get copied, the cell uses an enzyme called DNA polymerase to make the copy. This enzyme is not able to get started without the cell providing a starting point made from RNA called an RNA primer. This short sequence of RNA binds to the end of our chromosomes and DNA polymerase uses this to kick off replication, which it does all the way along one strand of the DNA double helix. On this strand, the section at the start that is bound by the RNA primer is never copied, so the whole thing gets slightly shorter. On the other strand (as DNA has two strands) the DNA polymerase has a more complicated process in order to copy the DNA sequence. DNA polymerase is not able to run along in the opposite direction in quite the same way, so it has to copy the second strand in chunks. In a similar way to the other strand, DNA polymerase needs help to get started in each of these chunks of replication, which it gets from small pieces of DNA called Okazaki fragments, that bind to the strand and help it to get started replicating. Also similar to the other strand, the last section of the DNA does not get replicated as there is not enough space for an Okazaki fragment to help DNA polymerase start working.

Surely this is a disastrous flaw in how our cells work? Big chunks of our DNA getting lost every time a cell divides? Fortunately for our body, we have a defence against this potential catastrophe: telomeres. Telomeres are a little like aglets, those metal or plastic caps on the end of your shoelaces. They help protect against the harm this degradation of our DNA could cause. At the ends of our chromosomes, telomeres are long sequences of TTAGGG in our genetic code, repeated hundreds or even thousands of times. These telomeric sequences don't code for any proteins and seem to have no other function other than to act as a sort of protective coating for our chromosomes. This means that every time our cells divide, a small number of those TTAGGG sections of the DNA is not copied and is lost, leaving the telomere slightly shorter, but otherwise not damaging the cell at all.

You may be asking yourself what this has to do with our body aging? The problem arises when those telomeres are eventually worn away so much that they can no longer protect our DNA effectively. When this happens those cells can no longer divide as they are not able to make full copies of the genes that they need to function properly. This effectively puts a limit on how many times the cells in our body can divide and this is thought to be a major factor in why our bodies age at a cellular level. As time goes on the number of cells in this state, known as senescence, in our bodies increases. These cells not only stop dividing, they also affect the way that they function, which genes they express, their rate of metabolism and even how they look. This has a knock-on effect on the tissues made up of these cells, with tissue function more impacted as the number of senescent cells increases over time. This increase makes our tissues degrade and impacts heavily on their ability to function.

There is some hope for potentially arresting or even reversing this destruction of our body's protective telomeres however: an enzyme called telomerase. This enzyme is present in our bodies, though it is only active in a small proportion of our cells. In particular, it is active in the cells that give rise to eggs and sperm, as well as some adult stem cells. Telomerase is able to extend the telomeres on chromosomes, essentially giving back the protective coverings that have been worn away from the end of their chromosomes. This enables these cells to go through the necessary large number of divisions they require (in the case of stem cells) and to produce eggs and sperm with the full-length telomeres that any new body they result in would need to grow and develop. Excitingly, this offers a potential target for future gene therapies directed at aging and age-related illnesses. It is also potentially useful in the treatment of some cancers, as restricting the activity of telomerase could put a limit on the uncontrolled cell division that gives rise to tumours. Interfering with telemorase expression, however, has been shown to induce cancer in some studies, so this fascinating and complex system needs a lot more research before treatments can be developed.

Erosion of our telomeres as our cells divide is just one of the reasons we age, however, and the cells of our body can age in other ways. One such way is down to simple clerical errors. It is again to do with our DNA and how it copies itself. Earlier in this chapter I mentioned how our chromosomes get copied and how these copies are almost exactly the same as each other, but not quite. DNA polymerase is able to copy the DNA strands with astonishing accuracy. In fact, some estimates have errors as low as one in every ten billion nucleotides (a nucleotide is one of the C, T, A or G bases in the genetic code linked to the phosphate group that join together to form the DNA strand). These errors are kept so low because not only does DNA polymerase assemble the new DNA, it also checks its work in a process known as 'proofreading'.

Why Does Our Body Age?

In other organisms, like viruses, that do not have this proofreading process, errors in DNA copying are far more prevalent. This is part of the reason why viruses are able to evolve into new variants and strains so quickly, and why we are able to catch new and exciting colds every year. While this increased speed of evolution might seem like a brilliant idea, mutations are extremely rarely a positive thing. For the vast majority of mutations, the impact is slightly negative or makes no real difference overall. Some can be catastrophic and lead to cancer, but many more just very slightly make things worse in the cell. Over time these small problems add up and the cells can enter a senescent state. There is some evidence that our cells enter into senescence as a way to prevent further issues leading to cancer, so aging tissues is seen as a much better choice than potential tumours. So over time, our cells accumulate more and more errors in their DNA, causing them to stop dividing. If it ended there, then it wouldn't be so much of an issue for our body. A slowly increasing number of non-dividing cells would not impact us so much. Unfortunately, senescent cells also send out signals that increase responses from our body, such as inflammation, which damages neighbouring cells and can cause them to enter senescence. Like a mouldy orange in the fruit bowl of our body, the spread of senescence in the tissue is strongly implicated in aging and age-related diseases. While this news of our bodies inevitably decaying may worry you, the main thing is to not stress about it as another factor in the aging process is actually stress. Psychological stress has been very strongly linked to an acceleration of the aging process. Being under prolonged stress causes the body to enter a sort of emergency state, which starts what is known as a cytokine storm. This is a response of our immune system, common when fighting viral diseases, where it releases large amounts of cytokines. Cytokines cause our body to fire up some of its defences, including helping to trigger the inflammation response. When we are sick or injured, the inflammation response can help protect our body and lead to a more efficient response against disease-causing germs. Unfortunately, prolonged elevation of cytokines and other stress-induced factors in our bloodstream can lead to the key drivers of aging that we've already discussed: celluar senescence and telomere shortening. This is unfortunately true of 'stresses' other than psychological stress. Exposure to anything that could damage our DNA, such as UV radiation, toxins inhaled while smoking, and even the highly reactive oxygen we use to power our body accelerates aging in our cells.

Well, this is all very grim reading for our bodies. An inexorable slide into decay, where our treacherous cells damage their DNA, destroy their protective telomeres and then recruit other cells to join them in their senescent state. Worst of all is, there is nothing we can do about it. Luckily, that's not exactly true. Research on aging and age-related diseases is an active and exciting area of biological research, with breakthroughs in recent years offering hope and optimism for the future. It's true that we can't stop time (yet), but there is evidence that we can reduce the impact of cellular aging and slow down the aging process. One area of research that shows some potential is where our bodies clear out senescent cells. Studies in mice, where trials of certain drugs known to encourage the clearing of senescent cells, have shown delays in the onset of certain age-related diseases. While promising, these results should be treated with very cautious optimism. Despite being mammals, our bodies have great differences from those of mice and this is a long way from a human cure, but research like this offers long-term hope for the treatment of age-related illnesses.

One thing is sure, with more research into new pharmaceuticals and other approaches, such as genetic therapies, we may have a chance in the future to help keep our bodies in better condition as we grow older. And who knows, maybe one day our bodies could be more like a jellyfish, at least in one important way.

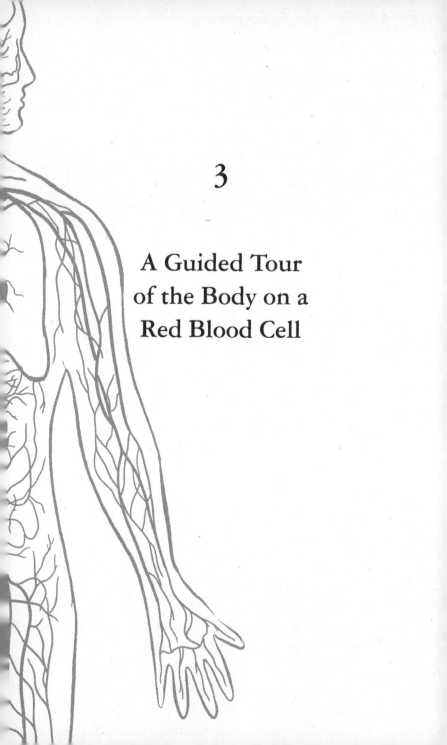

3

A Guided Tour
of the Body on a
Red Blood Cell

Our bodies have to overcome a lot of challenges in order to function in the way that they do. The simple act of keeping our bodies alive is actually far from simple, they are just so demanding. They need to be the right temperature; a change of just a few degrees causes catastrophic damage. Our bodies need food for energy, waste products removed, air to breathe, the list just goes on. Indeed, even keeping human cells in a dish in the lab alive requires maintaining them in very exact conditions. For some creatures this whole business of staying alive is much more straightforward. Take the amoeba, for example. Consisting of a single cell living in a watery environment, amoebae have an elegant solution to a lot of these demands from their tiny bodies; they just move. If they need food, they move around until they encounter something delicious and they engulf it inside their bodies by surrounding it with their cell membrane. To remove waste they do the opposite, essentially leaving the waste behind when they move off. Need oxygen? It readily diffuses through their membrane to where it's needed. As bodies get larger and more complex however, this becomes more difficult. These thicker bodies need to develop systems to transport things into, out of and around the body. Insects get oxygen into their bodies through tubes that link to the outside world called spiracles, though this process still relies on diffusion in a very similar way to amoebae. Indeed, it is this that limits the size insects can reach as more massive insects would not be able to get sufficient oxygen to all of their cells. This is because there are limits to how far oxygen in the air can diffuse through bodies, so the laws of physics prevent insects from growing larger.

Our bodies have different systems in place to get over these limits, and in particular there is a transport network that allows every cell in our body to receive and remove everything that it needs to keep our bodies going: the circulatory system. As the name suggests, the system circulates throughout our body, transporting what's needed and not needed in our blood. To get a feel for how this system works, we are going to join one of our blood cells on this round trip, and it's one of the cells we need to transport much needed oxygen: a red blood cell. So, if everyone is ready, it's time to hop onto our red blood cell. All aboard!

We start at our body's blood cell factory – the bone marrow. In an adult body there are not very many stem cells left, but the bone marrow is full of very active ones. These cells produce all of the different cell types found in our body's blood (red blood cells, white blood cells and platelets) at the truly staggering rate of approximately two million cells per second. Red blood cells are an unusual shape for a cell, they are what is known as a biconcave disc. They look like a ring doughnut where the hole in the middle didn't quite go all the way through, but there is still a distinct raised ring around the edge. This shape is believed to greatly increase the surface area of the cell, which is important for its primary function, carrying oxygen to and carbon dioxide away from almost every cell in your body. As we mentioned with large bodies above, having a long distance to the centre of the cell would decrease the efficiency for these gases being absorbed by the red blood cell. Another notable feature of red blood cells is that they are, well, red. This is what makes our blood red, and it is from a chemical that red blood cells are packed with: haemoglobin. Haemoglobin is a protein that contains iron, and it is this iron that enables it to bind to oxygen. Each haemoglobin is able to bind up to four oxygen molecules, and when they bind the haemoglobin changes colour to a brighter red. This is why arterial blood that is highly oxygenated is brighter in colour than deoxygenated venal blood.

Other than their distinctive shape and colour, red blood cells have one very distinctive feature. Or rather they are missing one; red blood cells have no nucleus. For most cells, not having a nucleus would be catastrophic, but as they mature, red blood cells lose theirs and it doesn't seem to impact on them too much.

This is because the DNA in their nucleus is no longer needed. Other cells, including white blood cells, need to produce proteins to carry out their functions, grow and divide. Red blood cells don't do any of these once they're fully mature, they are only needed for transport. It has also been hypothesised that having a nucleus would just take up space needed for haemoglobin and that squeezing through tiny capillaries is more difficult with a bulky nucleus, so maybe they're better off without one.

This quite unusual cell is mature, has all the proteins it needs and is ready to fulfil its purpose. To do that it first needs to make it to the engine of our body's circulatory system: the heart. The blood cells made in the bone marrow enter into the blood stream through the walls of tiny blood vessels throughout the marrow called capillaries. It follows the flow, alongside millions just like it, until it enters a major blood vessel, the vena cava. The flow of blood brings it from here into the heart. Our body's powerhouse, the pump that keeps life going, is quite literally its beating heart. Independent of any thought, our heart beats billions of times in the average body's lifetime. Four chambers make up the heart, two atria above two ventricles. Our blood cell enters the right atrium in a fairly steady flow of blood, but when the atrium contracts things get much more rapid. The blood is squeezed into the right ventricle in a swirling swish, before the ventricle itself contracts and our red blood cell and its sister cells are pushed out of the heart at speed into the pulmonary artery. This is the only artery where our blood is dark and deoxygenated, though it won't be like this for long as the next destination is our body's centre for gas exchange: the lungs.

It's common when teaching students about breathing to use empty bags to represent the lungs, showing how they expand on inhaling and deflate on exhaling. While useful it doesn't give a good representation of how amazing our body's lungs really are. Inside our lungs is a complex network of branching tubes that lead to tiny sacs called alveoli. There are millions of these alveoli in each lung and they massively increase the surface area of our lungs, for an adult body this is approximately the size of a tennis court. This huge surface area greatly increases the rate of gaseous exchange (oxygen in and carbon dioxide out) that our body can manage. Each alveolus has a lot of capillaries flowing around it, and it is to one of these that our red blood cell travels, and oxygen from the air in our lungs binds to the haemoglobin.

Our newly bright red-blood cell then continues its journey, this time heading back to the other side of the heart. It enters the left atrium, which contracts, sending it into the left ventricle. On the next contraction, our red blood cell gets driven into the largest blood vessel in the body: the aorta. The aorta comes under huge pressure from the flow of blood coming from the heart, and its walls are very strong and elastic. They stretch and contract with the flow of blood, which helps maintain blood pressure between heartbeats. Our red blood cell is swirled and juddered around in this maelstrom until it leaves the aorta through one of the other arteries that branch off of it. Some of these lead down into the body, bringing vital blood supply to the cells of our organs, but our red blood cell takes a different path: upwards.

The pressure from our beating heart and the pulsing of the aorta keeps the blood flowing against gravity as it passes through the carotid arteries in our neck and on into the brain. As the red blood cell travels, the blood vessels get smaller and branch off in different directions. A number of other cells follow these branches to supply oxygen to different parts of the brain, but our cell continues on into the anterior choroidal artery. It now enters the part of the brain responsible for motor control and vision, and into tiny capillaries. Here the vessels are so small that the lack of a bulky nucleus may be the only reason it is able to fit. Once it reaches a cell that has a lot of carbon dioxide, a neat chemical swap happens. High levels of carbon dioxide cause water, and other liquids in the body, to become slightly acidic. This lower pH causes haemoglobin to release its oxygen, which can then diffuse into the very neurons in this region of the brain that allow you to see, read and turn the pages of this book. The firing of neurons to complete these actions requires large amounts of energy, which in turn requires a plentiful supply of oxygen and generates a lot of carbon dioxide.

The newly vacant haemoglobin in our red blood cell then binds some of the carbon dioxide and turns from bright red to dark red once more. Our red blood cell now leaves the brain and needs to go back to the lungs to release its waste carbon dioxide and pick up more oxygen. It leaves the small capillaries and flows down one of the jugular veins to the heart. From here it is pumped once more into the lungs, where the oxygen-rich environment causes an increase in pH and this makes our red blood cells' haemoglobin release its carbon dioxide and bind to oxygen from the newly breathed-in air.

This entire process of circulating the body takes about one minute to complete. Our red blood cell is far from finished however; it will survive for up to four months, carrying out its vital courier service for our body. Each one of the many millions of red blood cells will carry out its journey tens of thousands of times before they are worn out and removed from circulation at a rate of about five million per second. Our body's transport system is astonishingly complex, with many moving parts all powered by our hard-working heart and the cells it moves around every second of every day.

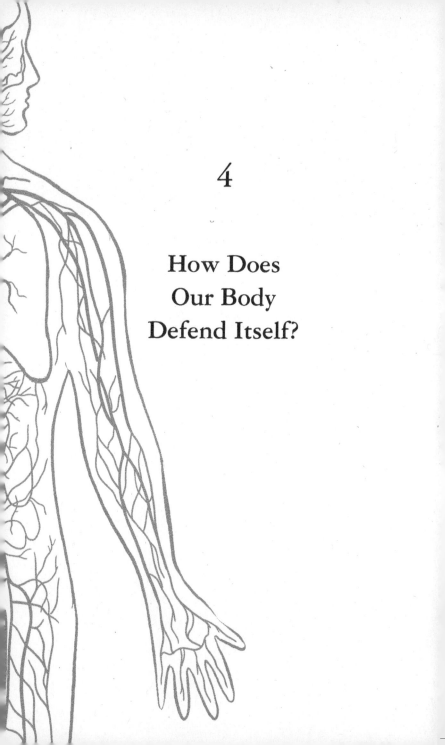

4

How Does Our Body Defend Itself?

One of the first things you learn as a biology student is that microbes are everywhere. The word used is 'ubiquitous' and a researcher often needs to go to great lengths to avoid the bacteria, fungi and viruses that are all around us. When we need to keep potential biological contaminants out, we use sterilised equipment, on sterilised surfaces and even use flames to prevent contaminations in the air from landing where we are working. Even with these precautions, every scientist has seen an unwanted colony of bacteria or a blue fuzzy fungus on their supposedly clean agar plate.

For any germophobes reading, this is probably horrifying and that is only when you consider the vast number of microbes in our surroundings. An even richer source of microbial life is the people around us. They are constantly shedding, exhaling, sneezing and coughing germs, not to mention leaving them behind on everything they touch. It is a hazardous environment outside our bodies, but it doesn't get much better on and in our bodies either. In fact, there are more bacterial cells than human cells in our bodies, and about ten times as many viruses. Luckily most of these are not pathogenic, so they do us no harm. Indeed, they can be quite beneficial, such as those in our gut that can bring all sorts of aid to our body, like help with digestion, while some research suggests they can prevent harmful pathogens from establishing and even impact our mood and mental health. A small proportion of these microbes are not 'friendly' though, they are downright dangerous. Their goal (if microbes can be said to have a goal) is to inhabit our bodies and use them to reproduce in vast numbers that can then spread to other bodies.

This rain of harmful pathogens would be enough to overwhelm our body long before we made it to adulthood if our bodies didn't mount any defence against it. Luckily our bodies have developed sophisticated mechanisms to identify, control and eliminate infectious agents. The first challenge facing any pathogen is how to actually get inside our bodies. Our skin is the first line of defence, and it is quite a hostile environment for an invading bacterium or fungal spore. When a germ lands on our skin it is faced with a wall, much like those faced by potential invaders of a medieval city. It is exposed to the elements and all of the friendly bacteria and fungi already established on our skin mean there is fierce competition for limited resources. This first line of defence for our body is remarkably effective, very few germs that land on it end up making us ill in any way. We can see how well our skin protects us when we get even the slightest break in it. Unless we take a lot of care, minor scrapes and cuts become overrun with bacteria and the cut gets infected.

Scratches and scrapes are not the only gaps in our outer walls however; there are gateways that lead into our body's more vulnerable interior. The mucosal surfaces of our body, i.e. the wet parts coated with lubricating mucous, are much better breeding grounds for pathogens and there are several on or near the surface of our bodies where germs can get access. Our bodies need food, water, air and sensory inputs to have access in order to function properly, as well as a need to get rid of waste products. Our noses, mouths, eyes, ears and other openings all allow germs to get into our bodies. Once they are inside the walls of our skin, other methods of defence are applied: our immune system.

Our immune system is incredibly sophisticated, consisting of millions of cells working constantly to protect our body. While it is highly complex, it is usually considered to be made up of two separate but collaborative parts: the innate and adaptive. The innate immune system works on anything that it detects as being indicative of an invading germ, while the adaptive immune system has to be trained to recognise pathogens which it can then target specifically. When pathogens manage to enter our body, they then must overcome the first line of our innate immune system: secretions. Our body produces a wide range of secretions to help with various functions, such as digestion and movement, but also to protect our vulnerable tissues from damage and attack. Mucous, gastric acid, tears, sweat and oil are just some of the substances covering and oozing from various parts of our body. While many have multiple functions, they are also involved in making it difficult for microbes and parasites to take hold in our body. Should they manage to get past these sticky barriers, a more active part of our innate immune system takes over. As well as billions of red blood cells, our blood contains vast numbers of white blood cells whose function is to attack any invading bacterium or virus.

In order to do this, the cells of our immune system have to be able to tell the difference between our own cells and those from outside the body. It does this by detecting some part of the pathogen as being foreign. This can be one or more of the proteins on the outside of bacteria, the genetic material of viruses or even the entire entity. These are called PAMPs or DAMPs, which stands for pattern or danger association molecular patterns. Once they are found, the cells of our innate immune system move into action. Some of the cells, phagoctyes and neutrophils move around in our blood like security guards on patrol and they engulf the offending germ by 'swallowing' it and destroying it inside themselves. Alongside these security guards, there are mast cells that act like an alarm system. Rather than engulfing foreign invaders, mast cells release chemicals, such as histamine that trigger an inflammation response. This swelling and increase of blood flow brings more white blood cells to the area, helping speed up the immune response. It is also part of the cause of allergic reactions where the immune response is triggered by something that it shouldn't be. As histamine is such a central part of this inflammatory response, a key method for dealing with allergies are antihistamine drugs. Mast cells also release other chemicals known as cytokines ('cyto' means cell), which help trigger inflammation but also alert other immune cells and attract them to the site of infection.

Sometimes our cells themselves can be compromised by invaders, in particular viruses. Viruses are particularly fascinating as they are so tiny (they are often more than ten times smaller than a bacterial cell), consisting of little more than a protein coat and some genetic material, and yet capable of causing so much damage to our body. Indeed, a friend of mine, an eminent virologist, once told me that asking if viruses are actually alive can be a contentious question. If a virus does manage to enter one of our cells, it makes the cell into a virus factory. The virus hijacks the cell's machinery to create thousands of copies of itself. This is clearly not something our body wants to happen and once a cell has become compromised it signals to the very appropriately named Natural Killer cells to put it out of its misery. The stricken cell puts out signals on its membrane to tell any passing Natural Killer cell that it needs to be destroyed. The Natural Killer cells then release enzymes and cytotoxic chemicals that break down the infected cell, while also releasing cytokines to recruit macrophages (a type of white blood cell that engulfs and destroys pathogens) to the area to help remove any surviving pathogens and clean up the remains of dead cells.

Some cells of the innate immune response are much less active in attacking foreign germs; they instead act like wanted posters. Dendritic cells are called antigen-presenting cells. This means they engulf germs or parts of germs and find pieces specific to that germ called antigens. The dendritic cell then races to our lymph nodes to 'show' this antigen to the other part of our immune system: the adaptive. These lymph nodes are the ones on your neck or under your arms that feel swollen when you are fighting off an infection, and that is because they are being called into action by dendritic cells. Once presented with the 'wanted poster' antigen, the adaptive immune cells, trained to recognise a specific antigen, now act like bounty hunters. They have an identified target and they hunt it down.

There are two types of cells in the adaptive immune system, known as B-cells and T-cells. B-cells produce antibodies: proteins that will bind only to the specific antigen of the invading germ, which prevent the germ from functioning properly. For example, if a virus has these antibodies stuck to its outside it will no longer be able to enter our body's cells to infect it and produce more viruses. The antibodies also act like a flag, marking the germ for destruction and clearance from the body. T-cells, when presented with an antigen, form into an army of specialised types. The cytotoxic T-cells attack the foreign invaders directly, identifying their antigen and releasing proteins that break down the cell membrane of the cell or cause it to self-destruct in a process known as apoptosis. Another type of T-cell is a memory T-cell. These remain in reserve in the body, pre-programmed by a previous encounter to rapidly recognise the antigen of the invader, ready to leap into action should the same microbe invade the body again. This is what enables us to become 'immune' to some diseases and prevents us from catching many illnesses more than once.

Obviously, being immune to things is far better than going through the process of fighting off germs. Indeed, for lots of common illnesses, it is the immune response we mount to fight off infection that causes many of the symptoms that make our body feel miserable.

Runny noses, fever, swelling and inflammation are all ways our body tries to make itself a hostile place for germs to live. Fortunately for us, we have developed a way to train our adaptive immune system and have it on high alert without having to be ill in the first place: vaccines. Essentially, vaccines work by introducing a non-harmful antigen for a particular disease causing agent, but without introducing the agent itself. Our immune system doesn't realise this, and it presents the antigen to our T-cells as if an infection is taking place. They produce memory T-cells which are then in place, ready and waiting, in case the strain of flu virus or whatever the vaccine was designed against should make it to our body.

While our immune system is amazing at protecting us from the vast number of potentially harmful things it's exposed to, it can also go wrong. For a lot of people this manifests as allergic reactions to what should be benign things that we encounter, such as pollen or peanuts. While these reactions may be severe, for most people allergies are an inconvenience and allergens can sometimes be avoided completely. For other people a misfunctioning immune system can start to attack something they can't avoid: their own body. Known as autoimmune diseases, sometimes our immune cells begin to attack what they are supposed to protect. The symptoms of these diseases vary widely, from fatigue and skin rashes to much more serious cases that can even be fatal, such as multiple sclerosis and Addison's disease. What triggers our body's immune system to attack the body itself is not fully understood, with genetic and environmental factors, as well as infections, all implicated. While for some bodies it can be detrimental, in a world where germs are ubiquitous, our bodies would not survive for long without an immune system.

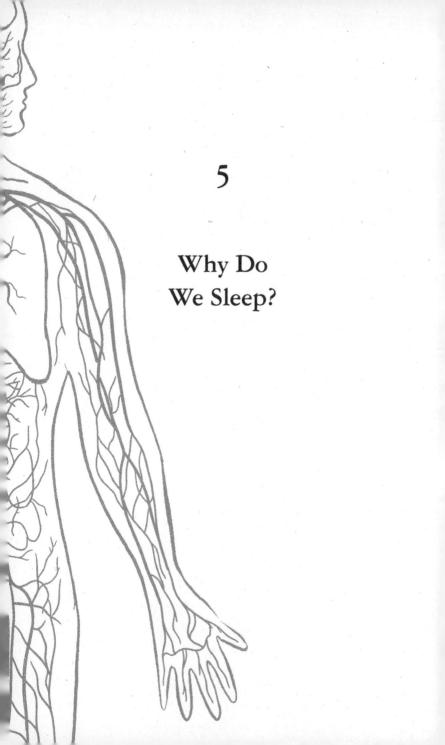

5

Why Do
We Sleep?

'Imagine it, please. The human bean who says he is fifty has been fast asleep for twenty years and is not even knowing where he is! Not even doing anything! Not even thinking!'

'It's a funny thought,' Sophie said.

'Exunckly,' the BFG said.

— *The BFG* by Roald Dahl

While he says it in a rather unusual way, it is hard to argue that the BFG doesn't have a point. We 'human beans' spend a large proportion of our lives asleep. Our bodies seem to shut down for hours at a time, using up a significant proportion of our lifespan and leaving ourselves completely vulnerable. In general, our bodies have evolved to remove traits that are wasteful and especially to remove traits that make us less likely to survive. While the chances of our body being under threat or attacked while we sleep are now greatly reduced, other animals living in the wild still need to shut down as we do on a regular basis. Even aquatic mammals like dolphins and whales sleep, though they have a different approach. As they are conscious breathers (i.e. they have to actively breathe, it isn't automatic like humans), whales and dolphins cannot become fully unconscious, and half of their brain sleeps at a time. In fact, almost all animals we have studied sleep in some way, including species as diverse as fruit flies and jellyfish. It seems that sleep is deeply embedded in how our bodies function, and despite the drawbacks of lost time and potential danger, it has remained through millions of years of evolution. Seeing as it's here to stay, in this chapter we will discuss what happens when we sleep and, crucially, why we need to do it at all?

We spend about one third of our life sleeping and it's important for our body that we spend the right third of each day in our bed. As diurnal creatures, we have evolved to function best during the day. Our eyes do not have the adaptations that nocturnal creatures, such as cats, have to see in the dark, instead they have evolved sophisticated colour vision that is only possible in brighter conditions. This means we are very good at spotting ripe red fruit to feed our bodies, but not so good at seeing in low light, so if we are going to spend a lot of time asleep then night-time is probably the best time for that. If going to sleep and feeling tired was a conscious decision then we could regulate that ourselves, but our body and in particular our brains control what is known as our circadian rhythm. This is our 'body clock', a pattern of processes that occur in a 24-hour cycle. What we eat, our physical activity and the environment can all impact on our circadian rhythm, but a key factor that impacts sleep is our exposure to light and dark.

When light enters our eyes, it reaches the retina at the back of the eyeball where the visual information is changed to electrical impulses that travel along our neurons to the brain. This is how we are able to see, and is a vital function of the eyes, but this information also goes to another part of our brain, the hypothalamus and more specifically a small region of the hypothalamus, the suprachiasmatic nuclei (We will call them SCN for short.). These two tiny structures are only about half a cubic millimetre in total volume, but they play a huge role in how our bodies regulate their day and are like the master clock of our body's daily rhythm. Within the cells of the SCN there are a series of genes known as 'clock genes' that interact with each other in complex interconnected feedback loops, but all you need to know is that these genes act as the 'tick' of our body clock. Left to themselves these genes would work away in a regular pattern, but they are subject to what is happening outside and within our body. Inputs to our SCN can alter our circadian rhythm by impacting on the expression patterns (when these genes are 'turned on' or not) of these clock genes. In essence, our body uses inputs, including light stimuli from our retinas, to set the time on our circadian clock. It's not just the amount of light either, with the wavelengths of light that occur more often as the sun is going down triggering more of a sleep response than those from the blue and green part of the spectrum typically seen earlier in the day. Until relatively recently this system helped our bodies become sleepy as night approached, but the advent of modern lighting technology and backlit screens in recent centuries has caused problems with our body clock. Eventually though, even our smartphones cannot keep us awake anymore and our body clock tells us we need to rest.

Once our body clock decides it's time for us to sleep, it starts to prepare us. As any parent knows, a bedtime routine is the best way to get a child to sleep and our body has its own routine to wind down for the night. When our body clock approaches bedtime, the SCN sends signals to the nearby pineal gland which begins to produce the hormone melatonin. Melatonin acts as a natural relaxant, putting our body into a state of 'quiet wakefulness', which is the calm feeling most people feel before they drift off to sleep. At around the same time, cells in the hypothalamus and brain stem produce a chemical called GABA which reduces the activity of what are known as 'arousal centres' in our brain. These arousal centres are the parts of our brain that cause us to be alert and motivated to move, eat and perform other activities. Another chemical produced by the brain to lower the activity of arousal centres is adenosine, whose mode of action can be counteracted by caffeine, which is why coffee can help stop you feeling drowsy.

Once we do drift off to sleep, the level of activity in parts of our brain changes throughout the night. The thalamus, which when we're awake acts as a relay station for most of your body's senses, reduces its processing of sensory information and also largely stops sending it to the brain's memory centres. This has the effect of shutting out the external world and helps our body to remain asleep despite minor stimuli from its surroundings. Our body has now entered one of the stages of sleep that we go through in cycles throughout the night. The first one is what is known as non-REM sleep. Sometimes split into three stages, non-REM sleep begins with lighter sleep where your heartbeat, eye movement and breathing all slow down. Your muscles begin to relax, with an occasional twitch, and your brain activity slows down. After several minutes in this state, your body temperature drops slightly and eye movements slow right down, almost to a complete stop. Your body then enters a deep period of sleep, your heartbeat and breathing slow even further and this period is key to feeling less tired and refreshed upon waking.

The next stage of sleep is (as you may have guessed) REM, or Rapid Eye Movement, sleep. Commonly, this first occurs between one and two hours after first falling asleep. As the name suggests, the previously quiescent eyes start to move rapidly along with raised heart rate, blood pressure and breathing. During REM sleep our thalamus starts to become more active, sending sounds, images and other input to our cerebral cortex causing us to dream. Fortunately, our brain stem also becomes more active during REM sleep, sending signals that relax the muscles to prevent us from acting out our dreams. The majority of our dreams occur during REM sleep, though some may happen during the other stages. The purpose of dreaming is not fully understood, but it is thought that it has a role in the consolidation of memories. The length of time spent in REM sleep typically gets longer throughout the night, with the first cycle's REM period lasting approximately ten minutes, with the final one lasting up to an hour. Once our body leaves the REM stage it goes through the whole sleep cycle several times until it wakes.

So that is what happens when we sleep, but why do we, and so many other species, need to sleep at all? There is no single answer to this question, though there are a lot of things our brain does during sleep that help it to function correctly. We have already discussed dreaming and its likely role in memory formation and consolidation. It would surely be hazardous to dream while we are moving around in the world, so being asleep during this process makes sense. It is also likely that prolonged periods of relative inactivity in our brain allow our body to perform maintenance of this vital organ. Waste products that build up during waking hours can be removed, hormone secretion can occur, and our neurons are free to communicate with each other. There was even a study in 2019 that showed pulses of blood and cerebrospinal fluid 'bathing' the brains of sleeping test subjects. This phenomenon is not fully understood but is believed to be part of the process that removes harmful substances and proteins from the brain. All of these processes may not be able to take place in a busy, working, wakeful brain and so our body might need to be asleep regularly to keep our brains healthy.

Whatever the reason, it is clear our bodies do need to sleep. While the longest period recorded without sleep was over eleven days, most of us suffer severe consequences long before this. Most adults need between eight and nine hours of sleep a night, though this can vary quite a lot. Children and teenagers usually require more while older people tend to make do with less. However much you need, if you don't get enough, it can very quickly impact on your ability to function and even damage your long-term health. Many of us have pulled an 'all-nighter' and know the drowsy feeling it brings being awake for twenty-four hours. You may find it more difficult to concentrate and being awake for this long can cause similar cognitive impairment to someone who is at the legal blood alcohol level to be allowed to drive. If you happen to push further, after thirty-six hours things like reaction time and attention span are significantly decreased, while the likelihood of drifting off into a microsleep (a very short involuntary sleep) without realising increases. After forty-eight hours without sleep you will have severely impaired judgement as well as a weakening of your immune system. Should you ever go three days without sleep your brain will begin to struggle to function properly, with mood swings, heightened anxiety and even hallucinations likely. Most sleep deprivation studies rarely go past three days due to ethical concerns for the test subjects, but past this stage anyone still deprived of sleep will mentally struggle even further. Basic cognitive processes will decline and your body's sense of what is real and what is delusion will fail.

Short-term sleep deprivation can have severe consequences; it is the direct cause of thousands of car crashes every year for example, but from a physiological point of view we can easily recover from occasional lack of sleep. Chronic sleep deprivation can, however, have quite severe consequences. Possibly as an attempt to compensate and to gain energy to keep going, our body has hormone changes when we are deprived of sleep. The hunger hormone ghrelin increases while the appetite control hormone leptin decreases. This causes us to crave high energy food and sleep-deprived individuals have a much higher risk of obesity and Type 2 diabetes. The decrease in our immune system caused by a lack of sleep also leaves our body more at risk to some cancers as well as heart disease and higher blood pressure. The direct impact on our brain leaves people who don't get enough sleep more at risk of developing dementia as well as mental health issues, such as depression and anxiety.

Sleep is a truly fascinating process, and research has a long way to go until we truly understand the question 'why do we need to sleep?' In humans, our large, complex brains need sleep to function correctly, but insects like honeybees and fruit flies need to sleep for the correct function of their memory. Even microscopic nematode worms have been shown to sleep, possibly to help with their development as they grow. Whatever the reason, it's clear that animals (including humans) don't just need sleep to rest, sleep is a vital process to keep our body and mind healthy.

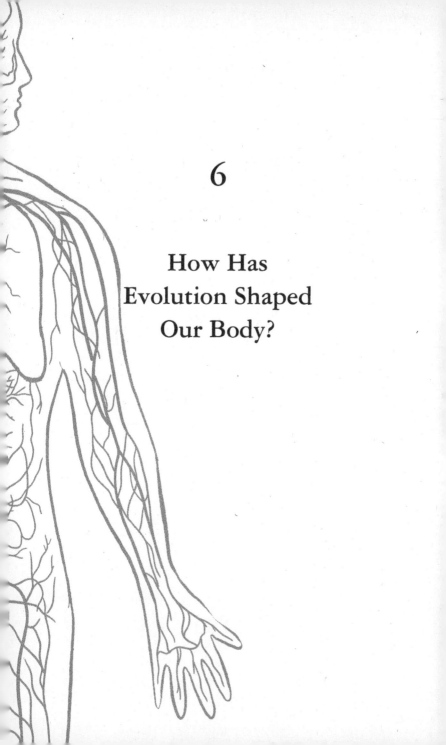

6

How Has
Evolution Shaped
Our Body?

If you walk on a beach after a particularly bad storm, you may spot some small creatures, blobs just a few centimetres long, washed up on the sand. These simple creatures are filter feeders known as sea squirts because, should you stand on one, they can squirt out water through their siphon. Early in their life as larvae they swim in the ocean, looking a little like a tadpole. Once they find a good spot to settle down, they digest their brain, lose their tail and a long fibrous structure called a notochord, and adopt their adult form. The adult then remains in place for the rest of the sea squirt's life, filtering food out of the seawater. That is until a wild storm stirs up the sea and causes it to wash up on a beach and be found by you.

You may well be wondering why this is relevant to how our body has evolved but, believe it or not, this sea squirt is a relative of yours. It is an admittedly distant relative; we diverged from them hundreds of millions of years ago, but we have some key things in common – most notably a notochord. When scientists like me study things, we love to classify them. Sorting things into different types makes it easier for us to understand and one of the main groups we sort species (which is itself another one of these groups) into is called a phylum. The phylum our bodies belong to is called the Chordata (after the notochord that all its members have), which contains all animals that have a backbone as well as small marine filter feeders like lancets and sea squirts. So, over tens of millions of years our bodies have evolved from simple tube-like filter feeders into something capable of baking a cake. How did that remarkable change happen and what signs of our simpler origins still remain?

It might be helpful to first understand how exactly evolution works in the first place. Evolution is not a completely directed process. There isn't a plan involved; it is triggered by random events. The DNA in our body's cells gets copied a vast number of times. Errors in this copying process are usually noticed and repaired, but a small percentage are not repaired, and a mutation happens. For the vast majority of these mutations, there is either no effect at all or there is a negative impact with many mutations leading to cancer. Every so often however, the change actually benefits the animal. If this benefit is significant and leads to that animal being more likely to reproduce than those animals without the muta-tion, then the new trait can become widespread and even lead to new species. Once you know how these changes occur, you need to understand the timescales involved. In complicated animals with long generation times like ours it takes an incredibly long time for noticeable change to happen to a species. Change through evolution can be sped up significantly by increasing what is known as evolutionary pressure. This occurs when there is an external factor that makes it more advantageous to have a new trait, and so it can become established in the population more quickly.

The classic example of this occurred during the Industrial Revolution in the UK. Peppered moths are small insects with white wings that are scattered with small black spots. This patterning makes them blend in remarkably well with the bark of trees in the forests they live in. During the Industrial Revolution, a sharp increase in air pollution due to the increased use of coal meant that many trees near built-up areas were coated in dark soot and ash. Peppered moths suddenly became much more visible to predators as their light-coloured wings stood out. The moth population always had variation in it, and the darker-coloured moths which had been a minority now became dominant in the more polluted areas. This increased pressure speeding up change applies to all evolving species, ourselves included.

In general, however, change has been slow, and we have evolved from our sea squirt cousins over many millions of years. There are some traces of that evolutionary journey in our body that you may not have even realised are still there. Due to the gradual nature of evolution, some things in our bodies are actually structures that have slowly changed form and function through the generations but are still there to remind us of our origins. When a sea squirt is filtering seawater to find food and oxygen, it pumps the water through its body using a muscular tube called a pharynx. This has slits in it that food is caught in known as gill slits, and these gill slits are one of the defining characteristics of all chordates, including ourselves. You may have gone back to re-read that last sentence, but yes, humans do have gill slits. In fish and other marine vertebrates, the gills are very obvious, in amphibians they are present in tadpoles and axolotls, but in mammals you need to look a little harder. The gill slits are no longer how we breathe, but they are present early in our development. The human embryo has gill slits that go on to form parts of our jaw and inner ear. They are still present in the adult form as the eustachian tubes that connect our mouth to our inner ear, and it is these tubes that we use to equalise pressure when our ears 'pop' on an aeroplane. We also retain the notochord and the nearby nerve chord seen in the swimming juvenile sea squirt, though they have now evolved into our spinal cord and the discs of cartilage between our vertebrae.

It's not just the things that have been with us from our simple marine ancestors either. Those much closer to us on the evolutionary tree have left their mark on our bodies too. One can be found in the corner of your eye; a small fold of the conjunctiva called the plica semilunaris. This is all that remains of the nictitating membrane, a third eyelid that closes horizontally across the eye, found in many species of reptiles, amphibians and fish. This membrane is also still present in some mammals, such as cats and polar bears, though in our body it has been reduced to a tiny vestige of what it once was. Human bodies also lack something that so many animals retain: a tail. Tails are extremely useful appendages, helping animals to balance, regulate temperature, acting as an extra limb and even as a handy seat if you're a kangaroo; so why do we not have them? Well in a way we do, there is in our skeleton a vestigial tail, a small remnant most people don't know exists unless they fall on their behind hard enough to hurt it. While I would personally love to have a tail, humans and apes' ancestors lost them around twenty-five million years ago. There is no consensus on what advantage this change brought, and it might just be a case of the change not leaving us any worse off so it never went away.

Our descent from apes has also left our body with some interesting remnants of life in the trees that we no longer use. The first is one that we only keep in the first few weeks of our life, the palmar grasp reflex. Human newborn babies are, while adorable, pretty pathetic compared to many other animals. Within minutes of hatching, baby turtles are able to dig their way out of the sand they were left in by their mother and cross the beach before immediately swimming in the ocean. Within an hour of being born many wild animal babies are able to stand and even run. Our babies pretty much lie there gurgling and sleeping. We do however have one extraordinary ability right away that we have inherited from our simian ancestors: the palmar grasp reflex. This reflex develops in the womb, when the baby can be seen grasping the umbilical cord, and for the first few months of its life our body can hold its own weight while holding onto a rod. This reflex means a baby can hold onto a bar or other narrow object and hang there for several minutes, something many adults including myself would genuinely struggle to accomplish. It is thought this reflex stems from the reflex of a baby monkey that hangs onto its mother's fur while she travels through the trees. Related to this reflex is another vestigial trait; the palmaris longus muscle. This is a narrow piece of muscle that runs from the elbow to the wrist. It is believed to have played a role in hanging from branches, but the ability to do so is not impacted at all in the approximately ten per cent of humans who do not have this muscle, so any advantage it once gave us is now lost. As it is now seemingly defunct, part of the palmaris longus is now commonly used as a graft material in reconstructive surgery.

Another possible remnant in our bodies can be found in our ears. When my dog is lying in her bed, there's something she does for an impressively large proportion of any day, and there's a sound outside. You can tell by looking at her ears. They perk up and move around, meaning she can find the direction the sound is coming from without making the supreme effort of actually lifting her head. The ears our body possesses are sadly quite stationary, though a small proportion of people can still wiggle them a little. The auricular muscles that control this movement are largely redundant in humans, though they are still present. In a similar way, we have a portion of our digestive system that is much larger in our evolutionary ancestors but has dwindled away in modern humans. It is a body part we all know the name of, though only because it frequently needs to be surgically removed: the appendix. Animals that eat a lot of leaves in their diet have a problem with digesting the cellulose that makes up a lot of plant tissues. This is such a problem that a whole group of animals, the ruminants, regurgitate their food to 'chew the cud' to help break it down. The digestive system of plant eaters also has a much larger appendix, which is believed to act as a store of bacteria that produce enzymes to help break this tough cellulose down. The appendix in our bodies no longer carries out this function; we are not able to break down cellulose like herbivores can. Instead, our much smaller appendix seems to play a role in our immune system, while also still acting as a store for gut bacteria that aids our digestion.

Whether evolution changes the function of parts of our body, makes them redundant or removes them entirely, it is an ongoing process. There is a strange belief that we have somehow 'completed' evolution, that we are at the top of the tree. This is the wrong way to look at this process that shapes our bodies. It is likely however, that the rate of evolutionary change has significantly slowed for humans. Unlike peppered moths during the Industrial Revolution, evolutionary pressure has decreased massively on humans in recent centuries. The majority of humans are not in danger of failing to survive to reproduce. Many of the cues that make us attractive to potential mating partners are social, and genetic factors impacting on who wins the 'survival of the fittest' are less important. This means that in several thousand years, if there are still humans, their bodies will likely be indistinguishable from our bodies today. It does not mean that there is no change however, and in a million years any humans that we would meet would likely have a very different body to the one reading this book.

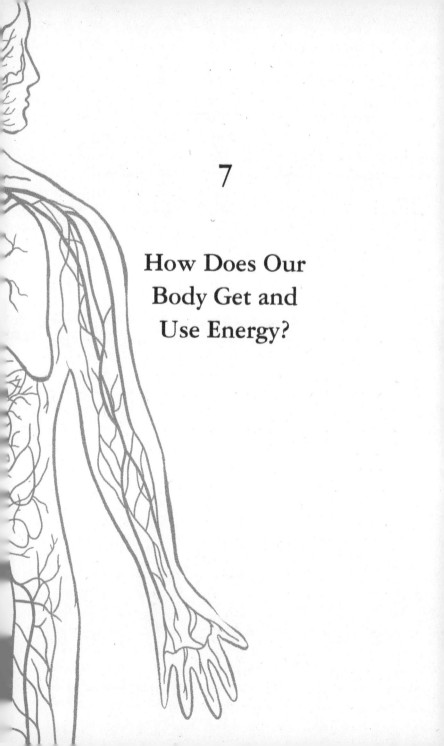

7

How Does Our Body Get and Use Energy?

One very important concept is central to understanding how our body and indeed the universe at large works. That might seem like a pretty strange statement; comparing our fairly tiny body to something as vast as the entire universe, but there are laws of physics that apply to the miniscule and the massive and everything in between. One of those rules is that there is a constant and never-changing amount of energy in the universe. That energy can change the form it takes, so the electrical energy in a battery is changed into emitted light energy when we turn on a torch, but it is still the same amount of energy. The first law of thermodynamics that we use to describe how this all works tells us that energy cannot be created or destroyed. So when we say that our body 'makes energy', that is not correct at all; we instead find a source of energy that we can then change into a form that we can use. Plants do this by harnessing the power of sunlight to make sugars from water and carbon dioxide, which can then be used as an energy source. As animals our bodies cannot photosynthesise our energy source in this way; we need to use other living things or the products they make in the form of food.

Modern society has allowed humans, especially in more affluent countries, to have access to a vast array of different foods. Technological advances in agriculture, refrigeration and shipping means that many of us are able to choose from more types and a greater volume of food than any previous generation. Even seasonality of foods that are only ripe at certain times of the year has little impact, with fruits and vegetables shipped or flown from thousands of miles away. We consume these foods in order to utilise certain components directly, such as calcium for bone strength and for use in our nervous system, and also for the energy stored in them to fuel all the functions that our body carries out every day. Some of those foods are more energy rich, and we measure that by how much heat energy they can produce which is measured in calories (*calor* being the Latin word for heat). But where exactly is that energy stored? It's not like a slice of toast is packed full of electricity, or an apple has batteries in it. The energy in food, and in pretty much all matter, is stored in the chemical bonds between the atoms that it is made of.

The cells in our body use the energy stored in a few different molecules as their ready source of energy, but for simplicity we will discuss the one that is used most of the time. It is called ATP, which stands for adenosine triphosphate. Essentially, the bond between the adenosine and one of the three phosphates is broken down, and this converts the energy of that bond into an energy that our cells can readily harness to use as it likes. Once the bond is broken, the ATP becomes ADP (adenosine diphosphate as it now has only two bonded phosphates) and to become 'recharged' this ADP needs to have a chemical reaction performed to bind another phosphate and make it into ATP again. This new ATP is now like a freshly charged battery for the cell to use. In order to power this chemical reaction to get fresh ATP, our body performs similar reactions to get energy that is stored in the chemical bonds of our food.

No matter what food we eat, in the end our body uses three different types of molecules as a source of energy to drive the chemical reaction needed to make ATP: carbohydrates, proteins and fats. Our body processes the three of these in slightly different ways, but in the end, they go through broadly the same two processes. These are aerobic and anaerobic respiration. As you may realise from the names, aerobic respiration uses oxygen and anaerobic doesn't.

If we follow the path of using a simple carbohydrate, glucose, to create ATP for cellular energy, the first anaerobic step happens at the start of the process. This happens in the gelatinous liquid that fills our cells, the cytoplasm. The glucose is broken down into two smaller molecules and breaking those bonds gains the cell enough energy to form two new ATP molecules. If the cell is short of oxygen, for instance if we have been strenuously exercising for a long period, then what remains of the glucose molecule is broken down and a further two ATP molecules can be produced. Interestingly, the waste product of this type of respiration in our body is lactic acid, which can build up in the muscles of endurance athletes. If our body is not short on oxygen however, aerobic respiration can produce up to thirty-six ATP from the same broken-down glucose molecule by moving the process into the 'powerhouse of the cell': the mitochondria.

Mitochondria are truly fascinating. They are tiny organelles in our cells that were once independent organisms. It is thought that very early on in our evolution, over a billion years ago, that our very early ancestors formed a symbiotic relationship with a bacterium that was very efficient at generating energy from food. This was one of the key moments in the evolution of life on earth as it helped to turn our cells from sluggish and inefficient layabouts to the dynamic machines they are today. They are inherited maternally, and have their own DNA, making them extremely useful in genealogical studies, but for our purposes here we are mostly interested in their ability to supercharge cellular respiration.

Once in our mitochondria, the broken-down glucose components are involved in a series of chemical reactions called the Krebs Cycle* (named after Hans Krebs who won a Nobel Prize for his work on it) which produces two more ATP and lots of molecules called NADH. NADH for our purposes can be thought of as an ingredient that is easily used to make ATP if you happen to have oxygen around. Embedded in the inner membranes of our body's mitochondria are a series of proteins. These are able to take the hydrogen (the H in NADH) from co-enzymes. This releases positively charged H+ atoms and electrons. The electrons are

---

* For time-saving purposes, and to not overwhelm you with dozens of complicated chemical formulae, this is a very high-level view of cellular respiration. The Krebs cycle involves lots more reactions than it would be possible to go through here, there are many more reactions involved across the whole process. And that doesn't even factor in the changes at the start of the process if our cells are getting energy from lipids or proteins. In the end however, the result is broadly the same. The first anaerobic step produces two ATP and some NADH, while the aerobic parts of the process provide up to thirty-six ATP. This is all from one glucose molecule, though those numbers are largely theoretical. Most cells will produce approximately thirty ATP as there are some inefficiencies depending on the cell type and other factors.

passed from one protein in the membrane to another in what is known as the electron transport chain. This series of reactions releases energy that is used to 'pump' the H+ ions across to the other side of the membrane. Building up the concentration of H+ causes a charge difference and a concentration gradient across the membrane. Whenever there is a higher concentration like this, it wants to equilibrate so there is a 'flow' from the higher concentration to the lower until it evens out. It is also driven electrically, as oxygen is an atom that likes to pick up negatively charged electrons making O− ions that the H+ ions are electrically attracted to, as opposite charges are attracted to each other. The H+ ions need a way across the membrane and that is provided by an enzyme called ATP synthase. This enzyme is sort of barrel-shaped and has lots of ADP and phosphorous molecules on it, and as the H+ ions pass through the barrel, that stream provides the energy to convert the ADP into ATP. Once the H+ ions combine with the O− ions they bond to form H_2O, so water is in fact a waste product of how we make energy.

While a lot of the ATP generated during this respiration is used to power processes in our cells, some of it is used in more obvious ways. One thing we all understand that uses energy is moving our muscles. But how exactly is ATP used to power our muscles?

Some muscle movements are involuntary and happen without conscious thought, and a good thing too as I would certainly find it hard to remember to tell my heart to beat seventy times a minute. If we want to say, move our eyes to read this book or stand up and move our arm to turn on the light if it is getting too dark to see the page, then our conscious mind gets involved. A part of our brain will initiate a process that sends an electrical signal along our neurons towards our muscles. This involves the movement of charged particles known as ions being let into the nerve cell by proteins that act like doors or gates opening up and allowing the ions to pass through. It is this flow of ions that we call electricity. Nerve cells are long and thin, with one end that has lots of branches (called dendrites after the Greek word for tree) at one end, a long thin middle section and then special communication sites at the end of the axon called synapses. The dendrites receive the signals and the flow of electricity goes along the axon until it reaches the end where the synapse consists of a gap between the axon and the dendrites of the next neuron. The flow of electricity cannot bridge this gap, so the neuron releases chemicals known as neurotransmitters that signal to the next neuron and cause it to start a new electrical wave in that cell.

This passing of the electrical impulse along a chain of neurons continues until the muscle the brain requires to move is reached. Here the synapse is not with another neuron, but with the muscle itself and is known as a neuromuscular junction. The stimulus of the electrical wave causes the neuromuscular junction to release a neurotransmitter: acetylcholine. When the acetylcholine binds to its receptor in the muscle cell, similar to what happens in neurons, an electrical impulse is formed. This time, however, it doesn't just cause the muscle to pass it along, the muscle takes this electrical impulse to mean it should do what muscles do best: contract. Our body has three main types of muscles; smooth muscles that are mostly for involuntary movements, such as those in our digestive system that move food along, cardiac muscles that are found in the heart and skeletal muscles that are the ones we actively control to move our body about. We will focus on the skeletal muscles here, which are made up of muscle fibres. When the electrical signal from our brain reaches a muscle in our body, it causes a series of chemical reactions that end up in the release of calcium ions into these muscle fibres. Inside each muscle fibre is an array of long, thin protein filaments. The majority of these are made up of two proteins, myosin and actin that occur in alternating bands throughout the muscle fibre. When the calcium released by the nerve impulse reaches these filaments, binding sites for the myosin filaments are exposed on the actin filaments. This causes the myosin sites to attach to the actin. The part of the myosin filament that attaches to the actin filament is shaped a little like a golf club, and it's the 'head' of this club that binds. Each of the heads has a store of chemical energy stored inside and it uses this energy to move while still bound to the actin filament. This causes the heads to pull the filaments along each other, which shortens the muscle fibre. In order to release from the binding sites, the muscle fibre releases ATP, which causes the heads to disengage by binding to the myosin heads itself. Once it is bound, the ATP degrades causing a release of energy that is then stored in the myosin head for the next muscle contraction. This degradation converts the ATP to ADP, which can then be 'recharged' to ATP to be used again.

Our body is made up of trillions of cells, each of which carries out a vast number of tasks and reactions that require energy. While we have the nice job of eating delicious food, our cells are incredibly busy using the energy from the bonds in those food molecules to provide our cells with usable ATP. It is this process that 'keeps the lights on' in our body, and in many ways is the fundamental biochemistry that keeps our body alive.

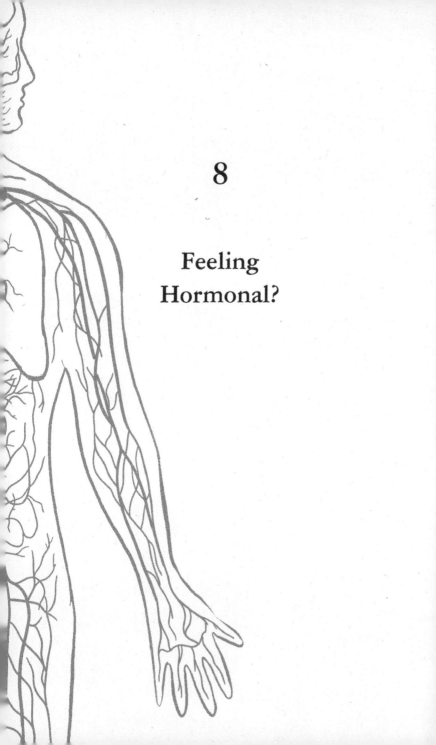

8

Feeling
Hormonal?

Hormones do not get the credit they deserve. The phrase 'hormonal' is usually accompanied by eye-rolling when a parent is describing their moody teenager or used in a somewhat patronising way to belittle the impact of the reproductive hormone cycle on a woman's mood. Hormones are so much more than something to make your body feel cranky and irritable though; hormones are central to almost every process in our bodies. They influence not just our emotions but also how and when we sleep, they control our appetite, our sex drive, growth, development, metabolism and, in general, help to keep our body in a steady state known as homeostasis. There are over fifty identified hormones in the human body, each with a very specific form and function. Without hormones our body would very quickly cease to function correctly, and in fact would never have been able to grow the body that is currently holding this book. So, while it is very clear that hormones are important, one key question we need to ask is: what exactly is a hormone?

Hormone is a fairly broad classification of molecules present in animals, plants and fungi. There are several different types of hormones based on what they're made of, but they are all used by our body in what is known as signalling. This means that our body produces hormones in a particular location, and they are then released so that they can cause a change to happen in another part of our body. In order to do this, hormones travel through our body (usually in the circulatory system) until they arrive at a cell that has the appropriate receptor. This works in many ways like a key and lock system, with the hormone 'key' only capable of fitting into the very particular cell type that has the right 'lock' to be opened. Once this occurs it acts as a signal to the cell to carry out a particular function. Thinking of hormones in this way, as molecules that travel through our body to bind to a specific receptor that causes a change of some kind, is of course oversimplified. To get a better idea of the hormones in our body, and the endocrine system that produces them, we will go through some of the best-known hormones and explore their extraordinary effect on our bodies.

Deep in our brain is a small yet highly significant gland called the pituitary gland. Despite being only the size of a bean, and weighing approximately half a gram, the pituitary is often called the 'master gland' due to its central role in so many aspects of our body's function and the endocrine system in particular. The pituitary gland secretes hormones involved in the regulation of body temperature, blood pressure and salt concentration, sexual development, pregnancy, metabolism and energy regulation, pain relief and even the function of other endocrine glands, such as the thyroid. It is also a key regulator of how our bodies grow. One of the hormones it produces is a long chain of peptides, the structural building blocks that make proteins, called somatotropin which is better known by its more informative name of human growth hormone. As the name suggests, human growth hormone is an anabolic molecule. Anabolic is usually something we associate with doping cases in sport, but it just means something involved in building things up. Human growth hormone (HGH) is vital to our body's development, repair, and growth. The pituitary gland releases HGH into our bloodstream in pulses, with the largest concentration linked to our circadian rhythm with larger pulses occurring during particular phases of our sleep cycle.

There are also gender and age variations, with, perhaps unsurprisingly, a marked decrease in HGH levels found as we get older and a noticeable increase in teenagers and young adults. When HGH is released into our bloodstream it causes a wide range of effects throughout our body, depending on which type of cell it binds to. For many cells it directly stimulates growth, while in others, such as liver cells, it causes a release of other growth factors that can cause our bones to lengthen. HGH can also change our metabolism, causing our body to use up fat and to deposit minerals into our bones. This wide range of functions mean that people who have HGH deficiencies display a wide range of symptoms, from reduced growth to a higher tendency of diseases like obesity and osteoporosis. People who have excess HGH may suffer from excess growth, oversized bones, such as the jaw and digits, and commonly develop tumours in the pituitary region of the brain. One major role of the endocrine system is to maintain homeostasis, and this includes for the hormones it produces. Too high or too low levels of hormones can be extremely damaging to our body, and sometimes hormones themselves help to regulate each other, such as the hormones involved in controlling our appetite.

The first law of thermodynamics is one of the fundamental laws of physics and if put simply it tells us that energy cannot be created or destroyed. Like the rest of the universe, our bodies obey this fundamental law and the energy we use on a constant basis has to come from somewhere, namely the food we consume. Earlier in the history of our species, our body's main problem was finding enough energy to maintain itself on a daily basis. It was likely that a persistent sense of hunger was simply part of early humans' life, so being hungry most of the time made a lot of sense from the body's point of view. On occasion though, it is very likely that there was an excess of food. Maybe a very successful hunt, or a glut of fruit in a particular season, would mean our technologically more primitive ancestors would have more food than they could possibly eat in one day. Maintaining a gnawing hunger and drive to eat in these circumstances would not be helpful to our bodies, in fact it could be downright dangerous. As we have evolved, and in particular due to recent advances in agriculture and food preservation technology, most humans live in a situation where the equivalent of a successful mammoth hunt happens every day. A stroll to the shops or a quick trip to the fridge enables us to access more calories than our body could use in a day of even quite strenuous exercise. For both our ancestors and ourselves, this windfall of food presents our bodies with a similar problem: when should we stop eating? Fortunately for our conscious minds, which are often quite wrong in answering this question (I know mine certainly is in the presence of a dessert menu), our hormones not only cause us to feel hunger they also tell us when we should stop eating.

When we metabolise sugar to get energy for our cells it is commonly in the form of a simple sugar called glucose. When we eat something, the levels of glucose in our blood usually go up, and the glucose is transported around our body for our cells to use it as fuel to make energy. As with everything to do with our body, homeostasis is vital and the level of glucose in our blood must be tightly controlled. If it goes too high for a prolonged period it can cause damage to our tissues, in particular the eyes, nerves and kidneys as seen in untreated diabetes. Should our blood sugar go too low, it doesn't just make us bad-tempered, it can lead to extreme lethargy and, if prolonged, to a coma or even death. Regulation of this vital issue of blood glucose levels is largely moderated by two hormones that work in opposite directions to maintain safe concentrations. When our blood sugar level starts to climb, our body releases possibly the most well-known hormone of them all, insulin. Produced in the pancreas, insulin is released on high blood glucose levels and initiates up to three processes to reduce the amount of sugar in our blood. The first is glycolysis. This produces ATP which is immediately used around the body as an energy source. The second is lipogenesis, which, as the name suggests, is the production of lipids. Lipids are then stored in the fatty adipose tissue around our body as a form of long-term energy storage. Lastly, insulin is able to start the process of glycogenesis. This is where the glucose, which is a very simple molecule of sugar, is converted into a much more complex type of carbohydrate called glycogen. This is then sent to other parts of the body, primarily the liver and muscle cells and glycogen acts as a short-term energy store. Glycogenesis is easily reversible, so the glycogen acts as a kind of battery for the body, that can be readily converted back to glucose to use as an energy source when required.

When insulin succeeds in lowering our blood sugar levels, or if we've not had anything to eat for a while, it is the turn of the other side of this hormone tug of war to come into action. Glucagon essentially does the opposite of what insulin does, causing glucose to be released from storage as glycogen and into the blood as well as causing the breakdown of other energy sources, such as amino acids and fats. Because their release is stimulated by high or low blood glucose, insulin and glucose also indirectly regulate each other, with the higher blood sugar induced by glucagon stimulating the release of insulin and vice versa. This enables our body to maintain blood sugar in a safe range, while also enabling us to store energy and use up those stores when we need to. This alone is a pretty nifty hormonal mechanism, but like most hormones there is more to it than that. Not only do insulin and glucagon regulate blood sugar and each other, they also impact on hormones that drive our hunger and satiety responses.

When we are hungry, it's often our stomach that we feel trying to bring our attention to this fact. We feel it gurgling away, rumbling and generally feeling like it is empty and we need to feed it. This is largely down to the action of a hormone produced mostly in our stomach called ghrelin. When we are low on food, cells in our digestive tract release ghrelin into the bloodstream. There is evidence to suggest that this production of ghrelin is stimulated by high levels of glucagon in the blood which occurs when blood glucose is low. When the elevated ghrelin reaches our brain, it triggers a hunger response from our hypothalamus. This causes our behaviour to change, and animals with raised ghrelin start searching more actively for food. It also causes our stomach to prepare for our next meal, so it produces more gastric juices, and it is this that causes the sometimes spectacular noises emanating from our bellies. Once we have eaten, ghrelin is also a key player in the reward system our body uses to encourage us to eat high energy foods. Also, after eating, our body releases insulin into our bloodstream, this insulin then helps to lower the concentration of ghrelin.

On the opposite side of ghrelin in this hormonal tug of war is leptin. Leptin is produced by the fatty adipose tissue stored around our body. These fatty tissues are essential stores of energy that our body uses in times of low food availability. So, fat can be a good thing but as we have discussed, keeping things in the middle is how our body works best. What our body needs to do is to keep our adipose stores at a certain level, and this means we need to control our hunger response.

This is where leptin comes in. Leptin production is partly determined by the amount of fat our body has in reserve, with an increase in body fat leading to higher leptin levels. It is also increased by the presence of higher insulin levels in the blood. When this happens, leptin causes our hypothalamus to lower our hunger response and to feel full.

These four hormones are only a small window into the wonderful and complex interactions of our endocrine system. So, whether it's these four hormones that are making you feel hungry or full, adrenaline and cortisol are giving you a rush of energy to deal with stress, melatonin is helping you drift off to sleep, thyroid hormone is regulating your energy levels, oxytocin is making you feel a bond with your new baby, or oestrogen, testosterone and progesterone are helping you make a new baby, just remember that all of these are you feeling hormonal.

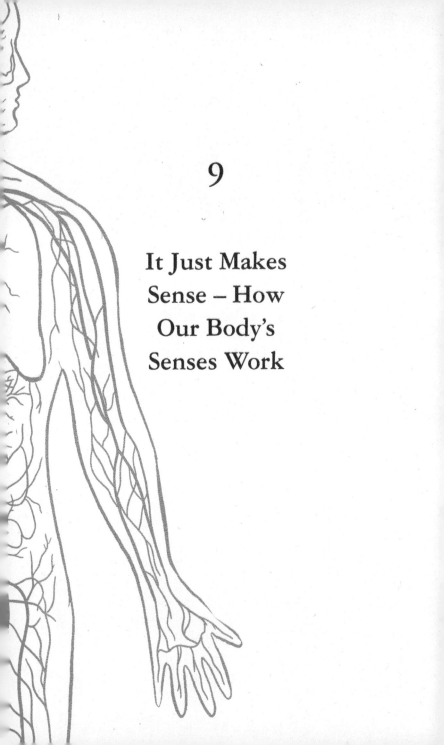

9

It Just Makes Sense – How Our Body's Senses Work

Our body is not kept in isolation. No matter where you happen to be reading this book, whether it's sitting on a train, lying in bed or perhaps it's the audio version and you're lucky enough to have me read it to you, your body is somewhere. And that somewhere has consequences, impacts and interactions with your body. Our bodies need to keep themselves safe, find food and water, search for a place to rest, interact with other bodies in various ways and a million other things that need to be navigated in our life. To do this successfully our body needs to have inputs from the external environment and from within itself so that it can act and react appropriately. These inputs come via what we broadly call our senses. Classically, we have five senses: sight, hearing, touch, smell and taste. We of course have many more senses than this with balance, hunger, thirst, pain, proprioception (an awareness of where our body parts are) and many others not making the standard list of five. The information provided by these senses allows our body to make informed decisions to deal with the complex world we live in. Writing about how they all work and interact with each other would require a book of its own, so for now let's discuss how the five 'classic' senses actually work.

As this is a book and you are reading it, we shall begin with vision. How does your body see the words on the page? If you think about it, it seems like magic that we can see at all. Alas, it is not magic just simple physics married to some elegant biology and it shouldn't surprise you that it starts with light. To be very technical, we don't actually detect an object itself when we see it, instead we detect the light that is reflected off it. When light hits an object some of the wavelengths are absorbed while others are reflected away. How much light is reflected and what particular wavelengths are not absorbed is what determines how bright it seems and also what colour it is. So, if an object absorbs all the visible wavelengths except red, then it will reflect the red light and appear red to us. Whatever the colours that are reflected, the next step to our body seeing an object is for it to encounter our organs of vision: the eyes. Our eyes may not actually be the windows to the soul, but they are the windows through which light gets into our body. The pupil is an adjustable opening into the eye, covered with a lens that focuses the light onto a layer of cells on the back of our eye: the retina. The shape of the lens means that the image formed on our retina is actually upside down. This is never corrected, and our brains merely interpret the upside-down image and it 'flips' the right way up in our consciousness. If you're thinking this through, you might be asking how does the information get to our brain to do this? The fluid-filled interior of our eyeball ends at the retina, so from this point it is not light that brings this message to our brain. That information needs to be converted into a useful form that our brain can process.

The cells in our retina that sense the light from the outside world are called the photoreceptor cells. These are split into two types: rods and cones. Rods are much more sensitive than cones and are more prevalent towards the outer edges of the retina. For this reason, rods are especially useful in low light situations and for peripheral vision. Rods are unfortunately pretty poor at determining colour, and this is the main reason why colours are so much harder to make out at night or in poor lighting. In order to see colour, we need to use the other, less numerous photoreceptor cells: the cones. Though they make up less than ten per cent of the photoreceptors in our retina, cones are able to detect different wavelengths of light, allowing us to see in colour. There are usually three types of cone cells: S-cones, M-cones and L-cones that are able to detect short, medium and long wavelengths of light respectively. These roughly cover the blue, green, and yellow/red portions of the spectrum. People with colour blindness are often missing or have a dysfunction in one of these cone types, while some people have an extra type of cone and have what is called tetrachromatic vision. This means that they can see more colours than most people and it is far more common in women than men, which may explain why some men don't seem to be able to distinguish between subtle shades of colour that the women in their life insist are obvious.

Regardless of if the light hits a rod or cone cell, the next step is the same. Chemical pigments in the photoreceptor cells react to light, causing a series of chemical reactions that triggers an electrical signal to be sent along the optic nerve, which links the eye to the visual cortex and occipital lobe in our brain. Each photoreceptor cell sends information about the light that it detects, so in many ways these are like pixels of an image on a computer screen. Each of these signals is analysed in our brain for various types of information, such as colour, movement, shape and outline as well as cross-referencing information from both eyes simultaneously for depth perception. How exactly this is all done in a tiny fraction of a second is still not fully understood, though it involves a huge amount of our brain to do so. Estimates of up to half of our brain tissue being involved in vision are common in those who study this, and considering how much visual information we process every day that is far from surprising.

While sight is a very dominant sensory input, there are four more of the classic senses to consider. Two of them are completely intertwined so it's best to talk about them together. Anyone who has ever had a heavy cold can probably guess which two: smell and taste. These two senses are closely linked as they process broadly the same information; detection of chemical substances. In the case of smell, this detection is of airborne particles from our environment entering our body through the nostrils. Higher up in our nose are millions of sensory cells called olfactory sensory neurons. These are stimulated by encountering particular chemicals, which then sends a nerve impulse to the brain. As odours are hugely complex mixtures of chemicals, many different olfactory sensory neurons are stimulated by any particular scent, and it is this combination of stimulations that our brain registers as a smell.

Our sense of taste works in a similar fashion, except the chemicals that are detected are not airborne, instead they are carried in a liquid: saliva. When we put food into our mouth, the first step to tasting it is for the food to be mixed with saliva. This starts the process of breaking down the food with enzymes found in our saliva, but it also enables us to taste our food. Our tongue is covered in hundreds of small lumps known as taste papillae. These papillae contain several taste buds and sensory cells that become stimulated by certain classes of flavours. Flavours are usually divided into sweet, sour, bitter, salty, and savoury or umami. Some of the sensory cells in our taste buds are only stimulated by one of the flavours, and they transmit the intensity of that particular flavour to the brain. The remaining sensory cells are able to detect all of the flavours to varying degrees, and they send back information about the mix of flavours they have encountered. Some of the chemicals in the food also become airborne, where they reach the opening to our nose found near the roof of our mouth. This causes our olfactory sensory neurons to become stimulated and send information to our brain, and it is this mixture of smell, flavours and intensity that allows our brain to know the taste of what we are eating. When we have a heavy cold, or for those who have lost their sense of smell, the lack of an olfactory stimulus when we eat can greatly reduce the intensity of taste. In matters of taste, it really is a combination of senses that counts.

In Ancient Greece, when people needed to know some information, they would often consult the oracle, a mystical source of knowledge. When our body needs to hear some vital information however, it instead consults the auricles, also known as our ears. When you look at our ears (at least our outer ears that are visible outside our body) they seem to have a bizarre shape. That shape has evolved to gather and direct the sounds that we need to hear into our middle and inner ears, that are inside our heads. The interesting thing is that unlike other animals, such as horses or cats, whose directable ears largely gather noise from one direction, our ears are able to gather sound from all around us. This gathering of sound is more efficient with sounds coming from the way we are facing, so this also helps us assess which direction the sound is coming from. Once our auricles have gathered in the sound and re-directed it into our ear canals towards our middle ear, their job is done.

In order to understand how the rest of the hearing process works we need to first realise what a sound actually is. Sound is a vibration that travels through the air or some other medium in the form of a wave. If you visualise this wave it looks like a line that goes up and down in a, well, wavelike motion with peaks and troughs. The height, or amplitude, of the waves determines how strong or loud they are. The distance between one peak and the next is the frequency, and the higher the frequency (i.e. the closer the waves are together, so they're more frequent) the sharper and shriller the sound. So, a whistle will have a high frequency while a drum will usually have a lower frequency. These soundwaves travel down our ear canal until they encounter a thin oval membrane called the tympanum or ear drum. Like a real drum, when the soundwaves strike the tympanum, they cause it to vibrate with the same frequency sending pressure waves into the middle ear. Here they meet the three smallest bones in our body, the malleus, the incus and the stapes (these are just Latin words for hammer, anvil and stirrup, which these bones are sort of shaped like). The pressure waves hit these bones which transmit and increase the amplitude of the waves, making them louder. On a slight side note, changing the amplitude of the sound waves like this is why this process is called amplification.

The last of the small bones, the stapes, strikes against the inner ear, transmitting the soundwaves into a fluid filled structure called the cochlea. This is shaped a little like a snail shell, and inside the fluid-filled interior is a membrane lined with sensory cells called hair cells. These hair cells move and bend when they encounter the soundwave running along the membrane and this movement causes channels to open up in the hair cells, releasing chemicals that start an electrical signal to be transmitted to the brain. Depending on where the hair cells are located in the cochlea, they are more likely to transmit these signals for different frequency sounds and this allows our brain to interpret the pitch of what we are hearing. These hair cells become damaged over time, leading to hearing loss in older people. Interestingly, those that detect higher-pitched sounds die off earlier, so even in early middle age our bodies can't hear the same range as we did when we were teenagers.

It is always a clever literary device to end a chapter on a touching note, and this chapter will be no different. Touch is usually the first sense our body experiences, developing early in the womb, and is thought to be the last sense we lose before we die. Touch is vital to our body's ability to find its way through the world, helping it to identify potential threats as well as being key to our development and social interactions. To make it easier to understand, we will split touch into three different aspects: mechanoreception (the ability to feel objects or pressure), thermoception (the ability to sense temperature) and nociception (the ability to feel pain).

When we pick up an object and can feel the shape of it in our hand, there are four different types of mechanoreceptors in our skin sending information to our brain. All four of these receptor types work in broadly the same way, with the physical distortion of the receptor caused when our skin touches something causing it to send a signal to our brain. Each type, however, specialises in a slightly different type of stimulus. Some are triggered to respond only by short stimuli, so they will react when we pull on our socks for example, but they'll stop sending signals once those socks are compressing our skin for a prolonged period. Others will continue to send signals as long as they are stimulated. All four have different thresholds of sensitivity, so some will feel a tiny bump while others only detect the broad size and shape of objects. As with some of our other senses it is a combination of the different types of sensory cells signalling the brain that allows us to build up an idea of what we are touching. Certain parts of our skin, the fingertips for example, have far more of the sensors that detect fine detail, meaning that we can build up a more detailed idea of the object we have grabbed hold of than if it sits on our shoulder or knee.

If we happen to grab something that is sharp however, we need our body to react differently. Should we pick up a cactus covered in sharp spines, burn ourselves or spill acid on our skin, then our nociceptors kick into action and we feel pain. There are different receptor types for mechanical, thermal and chemical pain. They are usually dormant, but once a threshold is reached, they activate and send signals to our brain. Finally, for thermoception, our body uses two different types of thermoreceptors to detect heat and cold. For both of these, the amount of signal they send to the brain depends on the level of stimulus. In extreme heat or cold they fire more frequently than in more moderate conditions. This allows our brain to interpret a range of detected temperatures.

All of our senses enable our body to survive in the complex environment we find ourselves in. While they detect a wide range of stimuli, all our senses and the organs associated with them convert information from the world into electrical impulses to send to our brain. It is our brain in the end that has to interpret those often very complex signals, and that astounding organ then does something truly amazing. It makes sense of our senses.

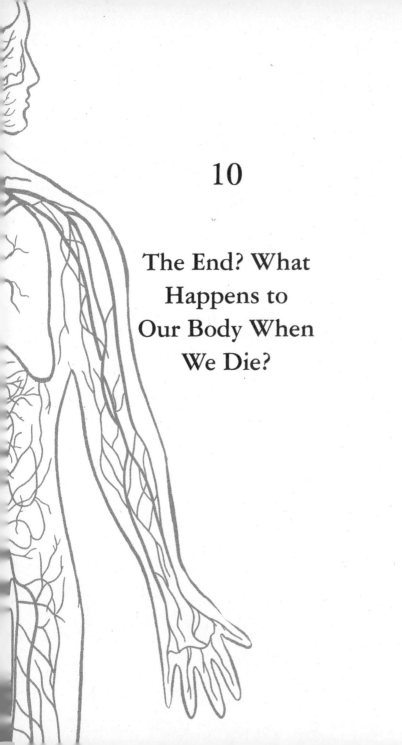

10

The End? What Happens to Our Body When We Die?

As I write this, there are an estimated 8.1 billion human bodies living on this planet (and thirteen in space on the International Space Station). As a species we have vastly different lives; our bodies are living in different places, seasons, and are spanning an age range of zero to almost one hundred and twenty years old. Those bodies look and act differently, with every body having its own thoughts, dreams and emotions. While we are all unique individuals, there are some things that all our bodies share, and one of those things is that we will all die one day. Unlike our friends the immortal jellyfish, a human body's lifespan is finite, and it will one day draw its final breath.

This is not the end for our body, however, that last intake of air instead signals merely the beginning of the end. Our bodies quickly begin to show obvious signs that they are no longer alive. One clear sign is that the heart is no longer beating. Signals from the brain stem that tell the heart muscles to contract are no longer being sent, so the vital flow of blood around the body ceases. If your body has pale skin tones, this lack of blood flow through its capillaries will cause the skin to become much paler. This process happens usually within half an hour, giving your body a deathly pale look and is known as *pallor mortis*. Soon after this, the blood cells will begin to accumulate in the part of the body that is closer to the ground. Gravity draws the red blood cells downwards causing a purplish discolouration of the skin known as *livor mortis*. While all of this is happening, the 'fire' of our metabolism going out will cause our body temperature to steadily decline. As well as making energy from food to fuel our bodies, the process by which we do this also generates heat. As metabolism ceases, this heat production slows down and eventually stops. This causes our body to gradually cool down at a rate of about 1°C per hour until it's around the same temperature as its surroundings in a process known as *algor mortis*. This will vary quite a lot depending on ambient temperature and the conditions the body is in, but it can be used as a rough approximation of time of death.

While the body is cooling down in this way, another much more famous sign of death begins, *rigor mortis*. Dead bodies are sometimes called 'stiffs' because a few hours after death the muscles can get locked, and the limbs stiffen. This is down to how our muscles work and the consequences of the energy source of the body, ATP, no longer being produced. When the muscle cells begin to degrade, a flood of calcium causes the proteins in the muscle fibres to lock together. Normally ATP is used to break these bonds and allow the muscles to relax, but as the oxygen supply needed for metabolism stops after death, the ATP stores in the muscles are quickly used up. The cells continue to make small amounts of ATP using anaerobic respiration on glycogen stored in the muscles, but after a few hours this too runs out. Unable to disrupt the bonds between the muscle proteins, the body thus becomes stiff. There is a widespread belief that *rigor mortis* lasts for long periods, but, in reality, the degradation of muscle cells causes it to end in less than a day in most cases.

While this is what happens to our body as a whole, many of our cells and tissues will continue on for some time, and, rather than a switch being turned off it is more like a fire that gradually goes out. Perhaps unsurprisingly, it is the parts of our body which need the most oxygen that stop functioning first. Top of the list for oxygen demand is usually our brain, which, outside of extreme exercise, uses about one fifth of the total oxygen we breathe in. The electrical and chemical signals constantly firing around our brain require very large amounts of energy and generating large amounts of energy in our body uses up a lot of oxygen. Without our lungs drawing in air and our heart pumping oxygen around in our red blood cells, the oxygen supply is suddenly cut off. The cells in our brain very rapidly deplete their supply of sugar and oxygen, and within a few minutes the cells begin to die. Within ten minutes huge numbers of neurons and other brain cells will cease to function. During this time frame, other highly demanding organs will begin to die. The liver uses vast amounts of energy while our body is alive. It is a major player in our metabolism, it helps regulate energy in our body and is a powerhouse of enzyme production. Like the brain this requires a lot of energy and oxygen, so the ending of that supply causes swift failure of our liver cells. Within half an hour of death a process called autolysis starts to occur.

When our blood stops flowing, it doesn't just stop oxygen being delivered, it also stops waste products being taken away from our cells. In particular, there is a rapid build-up of carbon dioxide which, being acidic, causes the pH in our cells to drop. This leads to the membranes inside our cells breaking down causing all of the organelles to release their contents into the cell. One particular organelle of interest here is called the lysozyme. This contains a large number of enzymes that in life the cell uses to break down things, but after death ruptured lysozymes break down the cell itself. In organs like the liver or pancreas where cells have large numbers of lysozymes, this process can be very rapid indeed. Across the whole body however, this inevitably happens, with cells releasing their contents as they burst. Autolysis doesn't just break down membranes, the enzymes released continue to break down all the constituents of our cells. Proteins, carbohydrates, fats, DNA and all the other components of our cells are broken down into the simpler building blocks they are composed of which is why this is known as decomposition. Almost all of the cells in our body begin to die off and decompose within a day or so of death, though some can last quite a bit longer. The more 'independent' white blood cells of our immune system continue for up to three days, still alive and able to carry out their basic functions. Eventually, even these will die off and our body is truly dead.

Or is it? At this point our body ceases to be a body as we have described throughout this book, but it does become an ecosystem. Estimates vary, but it is thought that about half the cells in our body are not actually human cells, but instead are microbial. Our microbiome consists of trillions of microbial (bacteria, viruses and fungi) cells that live in and on our body. There is substantial evidence that our microbiome can have positive effects on our digestion, immune system and even our mood. These microbes are only beneficial as long as our immune system keeps them in check and stops them from getting out of control. Once our body dies, that control stops. Added to this, the process of autolysis releases lots of broken-down nutri- ents, creating a wonderful environment for microbes to grow. And oh, do they grow! Within a very short time, our microbiome begins to break down our body while releasing all sorts of waste products and gases like ammonia and methane in a process called putrefaction. These gases cause the body to swell and bloat. The bacteria spread from the gut to the other soft tissues of the body, breaking organic material like amino acids down further, ending up with waste products with such charming names like putrescine and cadaverine. These waste products usually have very powerful and distinctive odours, and this is what we detect when we smell something rotten. From this point on, our body is merely a substrate for any insect, bacteria or fungus that can take advantage of this glut of nutrients we once called home. Eventually only those tissues that are mineralised or offer little by the way of nutrients, such as bone and hair remain.

That is pretty much the end of our body. The bustling ecosystem that our body provides upon death gradually quietens down as the nutrients we provided are depleted. If our body has been buried then the basic minerals remaining from the microbial and insect feast end up in the soil and enter the cycle of soil nutrients used by plants and animals. If the soil is acidic, then our bones will very slowly dissolve into the earth over a period of tens to hundreds of years. In very particular soil conditions this could even take thousands of years.

Eventually though, there will be nothing left, except maybe those fillings you got because you didn't floss.

I do not want to end this book on a low note though, and there is always time for looking on the bright side. Our bodies are truly remarkable. They grow in a truly astonishing way from one cell to around thirty trillion, are able to move, think, feel and change the world around us. Using those bodies, we have managed to achieve feats in science, art and culture that no other animal has achieved. We have used our bodies to compose music that our body is able to interpret to such a degree that it triggers joy or makes us weep. We have even sent bodies into space and back again. But if that isn't enough for you, if you think your body is not remarkable, you should consider what it's made of. The atoms that make up your body are widely varied: calcium, phosphorous, iron, oxygen and many others. Most of these elements can only be formed in the hearts of stars or in the vast surge of energy when a star goes supernova. So, no matter how ordinary you think your body is, you should remember that every body is, in a way, made of stardust.

Acknowledgements

I would like to thank everyone who has helped this book become a reality. Many thanks to everyone at Seven Dials and Orion Publishing, in particular, to Tierney Witty for helping this novice put together his first book. I'd also like to thank Dr Andrew Hogan and Loren Kell for their helpful input in areas they know so much more about than I do. Thank you to everyone at Bold Management for helping get this project running, in particular Felan and Jade who did everything else so I could get on with the fun job of writing. Thanks also to my university friends who were proofreaders and sounding boards. And finally, thanks to Joan, who helped me stop being too technical and always pointed me in the right direction. It wouldn't have been anywhere near as good without you.